Abdelhafid Mimouni

Spectrochemistry Decoded: Double Groups and Their Applications

Abdelhafid Mimouni

Spectrochemistry Decoded: Double Groups and Their Applications

ScienciaScripts

Imprint

Any brand names and product names mentioned in this book are subject to trademark, brand or patent protection and are trademarks or registered trademarks of their respective holders. The use of brand names, product names, common names, trade names, product descriptions etc. even without a particular marking in this work is in no way to be construed to mean that such names may be regarded as unrestricted in respect of trademark and brand protection legislation and could thus be used by anyone.

Cover image: www.ingimage.com

This book is a translation from the original published under ISBN 978-620-6-71630-3.

Publisher:
Sciencia Scripts
is a trademark of
Dodo Books Indian Ocean Ltd. and OmniScriptum S.R.L publishing group

120 High Road, East Finchley, London, N2 9ED, United Kingdom
Str. Armeneasca 28/1, office 1, Chisinau MD-2012, Republic of Moldova, Europe
Printed at: see last page
ISBN: 978-620-7-89766-7

Spectrochemistry Decoded: Double Groups and Their Applications

Author: Dr. Abdelhafid Mimouni is an independent researcher specializing in bioinorganic systems chemistry, with extensive expertise in macromolecular synthesis and characterization. He obtained his PhD in chemistry from the University of Paris XII in 1997 and a Diplôme d'Études Approfondies in bioinorganic systems from the University of Paris XI in 1993.

Summary:

Spectrochemistry is an essential scientific field that studies the interaction between light and matter at the molecular level. Through various techniques such as infrared, Raman, UV-Vis, NMR and other spectroscopies, it enables the detailed analysis of molecular structures, electronic transitions and molecular vibrations. Group theory plays a crucial role in classifying quantum states and predicting spectroscopic transitions based on molecular symmetries. This discipline has a wide range of applications, from materials characterization to environmental monitoring and pharmaceutical quality control. Constantly evolving, spectrochemistry explores new and emerging techniques and promises to continue pushing back the frontiers of scientific research and technological innovation, contributing to our understanding and impact on the modern world.

Table of contents

Introduction 3

Chapter 1: Theoretical bases of spectrochemistry 5

Chapter 2: Molecular spectroscopy 10

Chapter 3: Electron spectroscopy 13

Chapter 4: Advanced spectroscopy techniques 16

Chapter 5: Specific applications of spectrochemistry 19

Chapter 6: Future prospects and progress 27

Conclusion 30

List of symbols and abbreviations 32

References 37

Introduction

Spectrochemistry occupies a fundamental place in the physical and chemical sciences, offering invaluable tools for exploring the structure and interactions of molecules at the atomic and molecular levels. This interdisciplinary field combines the principles of quantum chemistry and spectroscopy to study how light interacts with matter, providing crucial information on the composition, structure and properties of materials.

Definition and importance in the physical and chemical sciences

Spectrochemistry can be defined as the study of the interactions between electromagnetic light and matter, probing the energy transitions and quantum states of atoms and molecules. In physics, it is an essential tool for characterizing the spectroscopic properties of substances, while in chemistry, it is indispensable for understanding the chemistry of reactions, molecular structure and chemical bonds.

Overview of modern spectroscopic techniques :

Modern spectroscopic techniques cover a wide range of experimental and theoretical methods, each adapted to specific aspects of molecular analysis. These include vibrational spectroscopy (IR), which analyzes molecular vibrations; rotational spectroscopy (microwave), which studies molecular rotation; electron spectroscopy (UV-Vis), which examines electronic transitions; and advanced techniques such as NMR and mass spectroscopy, used for structural analysis and materials characterization.

Book objectives and structure :

This book aims to explore in depth the theory and practical applications of spectrochemistry, focusing on the integration of group theory with the Schrödinger equation. Each chapter will address a specific aspect of spectrochemistry, starting with the theoretical foundations and developing through to practical applications in different scientific and technological fields. The aim is to provide readers with a comprehensive understanding of the underlying principles and practical tools needed to master modern spectrochemistry.

This introduction lays the foundations for an in-depth study of light-matter interactions and the advanced analytical methods used in spectrochemistry, while highlighting the importance of this field in scientific research and industry.

Chapter 1: Theoretical bases of spectrochemistry

Foundations of quantum mechanics

Quantum mechanics is the theoretical foundation of modern spectrochemistry. At the heart of this theory is the Schrödinger equation, which describes the evolution of wave functions in quantum systems. This fundamental equation integrates the concepts of quantum particles and their interactions, providing a precise mathematical description of quantum states and their associated energies.

Applying Schrödinger's equation to simple molecular systems enables us to model their energetic and structural behavior. For example, for a hydrogen atom, the Schrödinger equation predicts the discrete energy levels of the atomic orbitals, as well as the electron localization probabilities around the nucleus. This theoretical approach provides the basis for understanding phenomena observed in spectroscopy, such as electronic transitions and light-matter interactions.

Group theory

Group theory plays a central role in spectrochemistry, enabling the systematic analysis of molecular symmetries and spectroscopic transitions. Symmetric groups, such as double groups, are powerful mathematical tools for describing the symmetries and properties of molecular systems.

An introduction to group theory, itself a rich mathematical field, reveals how the symmetry elements of a molecule can be classified and exploited to interpret its spectra. For example, the symmetry operations of a molecule can be represented by group matrices, making it possible to predict selection rules in absorption and emission spectra. This theoretical approach is essential to distinguish between permitted transitions, which respect symmetry rules, and forbidden transitions, which violate them.

Concrete examples of the use of symmetry groups illustrate their application in spectral analysis. When studying molecular vibrations, for example, the symmetries of normal modes can be determined from representations of the molecular symmetry group, facilitating the interpretation of infrared spectra. Similarly, for polyatomic molecules, the use of group theory enables rotational and vibrational spectra to be predicted and analyzed with greater precision.

In summary, this first chapter lays the essential theoretical foundations for understanding spectrochemistry. By combining quantum mechanics with group theory, it provides a robust framework for the analysis and interpretation of observed spectroscopic phenomena, laying the foundations for further exploration of the practical applications of modern spectrochemistry.

Application of group theory to the Schrödinger equation for the hydrogen atom :

Schrödinger's equation is the mainstay of non-relativistic quantum mechanics, describing the evolution of the wave functions of quantum systems. For the hydrogen atom, this equation can be solved exactly thanks to the symmetries associated with the SO(3) group, which represents rotations in three-dimensional space. With this in mind, we explore how group theory facilitates the analysis and solution of the Schrödinger equation for this emblematic example.

1. Schrödinger equation for the hydrogen atom :

The hydrogen atom is a simple but fundamental quantum system, consisting of a positively charged proton and an electron orbiting around it. The non-relativistic Schrödinger equation for this system is given by :

$$((-2m/\hbar^2)\nabla^2 -(4\pi\epsilon_0 / r)e^2)\psi(r)=E\psi(r)$$

$\psi(r)$ is the wave function, dependent on position r.

- $^2 \nabla$ is the Laplacian operator, which represents the Laplacian of the scalar field ψ.
- mmm is the mass of the electron.
- $\hbar$ is Planck's reduced constant.
- e is the electron's charge.
- $^0 \epsilon$ is the permittivity of vacuum.
- r is the radial distance.
- E is the total energy of the electron.

2. Symmetry and symmetry groups :

The hydrogen atom has spherical symmetry, meaning that its physical properties are invariant to rotations in space. This symmetry is mathematically represented by the SO(3) group, whose elements are rotations around the origin.

3. Representations of the SO(3) group:

Solutions to Schrödinger's equation can be classified according to irreducible representations of the SO(3) group. These representations describe how wave functions change under the action of rotations in three-dimensional space. In physical terms, they characterize quantum states in terms of their orbital angular momentum and symmetry under rotations.

4. Solving the Schrödinger equation :

Using the properties of the SO(3) group, it is possible to express the wave functions $\psi(r)$ and energies E of the hydrogen atom's quantum levels. The atom's bound states, such as the s,p,d orbitals, correspond to different representations of the SO(3) group. For example, the s orbital is spherically symmetrical (trivial representation), while the ppp orbitals are axially symmetrical.

5. Wave function development :

The wave functions of the hydrogen atom can be developed in terms of spherical coordinates, reflecting the spherical symmetry of the atom.

Spherical harmonics and Laguerre polynomials are specific solutions of the Schrödinger equation for the hydrogen atom, deriving directly from the properties of the SO(3) group.

In conclusion, group theory, particularly the SO(3) group of rotations in three-dimensional space, plays a crucial role in the analysis and solution of the Schrödinger equation for the hydrogen atom. This approach makes it possible to classify quantum states according to their symmetry, thus facilitating the prediction and interpretation of observable spectroscopic properties, such as the selection rules for energetic transitions. A thorough understanding of these concepts is essential for applications in atomic, chemical and spectroscopic physics.

Chapter 2: Molecular spectroscopy

Vibrational spectroscopy

Vibrational spectroscopy, particularly in the infrared (IR) range, is a powerful method for studying molecular vibrations. It is based on the principle of energy absorption by a molecule as it passes from one vibrational state to another. The energy absorbed corresponds to the difference in energy between these states, and is typically measured in cm^{-1} (wave number).

The fundamental equation used to describe light-matter interaction in IR spectroscopy is based on Beer-Lambert's law:

$A=\epsilon cl$

Where:

- A is the absorbance,

- ϵ is the molar absorption coefficient,

- c is the sample concentration,

- l is the optical path thickness.

This relationship quantifies absorption intensity as a function of sample concentration and cell thickness.

Application example:

IR spectroscopy has a wide application in the identification of molecular bonds and conformations. For example, it is crucial in the study of polymers to determine their specific chemical structures and monitor polymerization reactions. In biology, it is used to analyze protein structures and molecular interactions. In environmental chemistry, it is used to monitor pollutant levels in air and water, thanks to the specificity of spectral signatures.

Rotational spectroscopy :

Rotational spectroscopy, mainly in the microwave range, focuses on the study of energetic transitions due to molecular rotation. This technique is particularly sensitive to molecular dipole moments and asymmetric molecular structures.

The rotational energy EJ of a diatomic molecule can be approximated by the expression :

$$E_J = J(J+1)\, \hbar\, /2I^2$$

Where:

- J is the quantum number of rotation,

- $\hbar$ is Planck's reduced constant,

- I is the molecule's moment of inertia.

This formula illustrates how rotational energy depends on the quantum number J and the moment of inertia of the molecule.

Application example:

In environmental spectrochemistry, rotational microwave spectroscopy is used to identify and quantify atmospheric gases such as carbon dioxide (CO_2) and methane (CH_4). These molecules exhibit distinct rotational transitions that are specific to their molecular structures. In industry, this technique is also used to monitor chemical processes and reactions in the gas phase, where precise detection of components is essential.

In conclusion, this chapter highlights the crucial importance of vibrational and rotational spectroscopy in the analysis of molecular properties. These techniques provide precise information on molecular structures, conformations and interactions, playing a central role in various scientific and technological fields, from fundamental research to industrial application.

Chapter 3: Electron spectroscopy

UV-Vis spectroscopy

UV-Vis (ultraviolet-visible) spectroscopy is a fundamental method for studying electronic transitions in chemical compounds. It relies on the absorption of light by a molecule's electrons, inducing transitions between different electronic energy levels.

Basic theory :

Light-matter interaction in UV-Vis spectroscopy is described by Beer-Lambert's law, which quantifies absorbance A as a function of sample concentration ccc, cell thickness l, and molar absorption coefficient ϵ:

$$A = \epsilon \cdot c \cdot l$$

Observed electronic transitions are generally classified into three main types: π-π^* (pi to pi star), n-π^* (non-bond to pi star), and σ-σ^* (sigma to sigma star) transitions, each corresponding to specific absorption energies in the UV-Vis region of the electromagnetic spectrum.

Application example:

In organic chemistry, UV-Vis spectroscopy is crucial for characterizing organic compounds such as dyes, polymers and aromatic compounds. For example, detecting the presence of unsaturated bonds (double or triple bonds) in a molecule can be done by observing π-π^* transitions in the UV-Vis spectrum.

In inorganic chemistry, this technique is used to analyze metal complexes and transition element compounds, where d- and f-electron transitions are frequently observed in the UV-Vis region.

Fluorescence spectroscopy

Fluorescence spectroscopy is a complementary technique to UV-Vis spectroscopy, exploring the light-emitting properties of molecules after they have been excited by an external light source. It relies on the phenomenon of non-radiative relaxation of excited electrons, which return to their ground state by emitting light of a longer wavelength than that absorbed.

Basic principles :

Fluorescence intensity I is proportional to sample concentration c, fluorescence quantum yield Φ , and optical absorption A :

$$I = \Phi \cdot A \cdot c$$

Fluorescent molecules have specific chemical structures that promote radiative relaxation of excited electrons, leading to characteristic fluorescence spectra with distinct emission peaks.

Application example:

In biology, fluorescence spectroscopy is used to study protein structure and dynamics, as well as molecular interactions in living cells. For example, fluorophores such as fluorescein are widely used to label specific biomolecules and track their localization and movement in biological systems.

In materials science, this technique is applied to characterize the optical and structural properties of materials such as luminescent polymers and fluorescent nanoparticles, providing valuable information on their composition and behavior in different environments.

In conclusion, this chapter highlights the fundamental importance of electron spectroscopy, in particular UV-Vis and fluorescence spectroscopy, in characterizing the electronic and luminescent properties of molecules and materials. These techniques play an essential role in many scientific fields, from chemistry and biology to materials science and technology.

Chapter 4: Advanced spectroscopy techniques

Nuclear Magnetic Resonance (NMR)

Nuclear Magnetic Resonance (NMR) is a powerful spectroscopic technique used primarily in structural chemistry to study the molecular composition and organization of substances. Based on the magnetic properties of atomic nuclei, NMR exploits the interaction between nuclear spin and an external magnetic field to determine molecular structure and analyze chemical interactions.

Fundamental principles: NMR is based on the phenomenon of nuclear precession, in which an atomic nucleus with a non-zero spin reacts to an external magnetic field B_0 by oscillating at a frequency called the Larmor frequency. For a nucleus of spin I, the Larmor frequency ω is given by the equation: $\omega = \gamma B_0$

where γ is the gyromagnetic ratio of the core.

When radio frequency (RF) is applied at resonance frequency ω, nuclei absorb energy and emit signals detected as NMR spectra. The intensity and position of these signals provide information about the chemical environment of the nuclei, including their atomic connectivity and molecular dynamics.

Applications: In structural chemistry, NMR is used to determine the three-dimensional structure of molecules, identify chemical bonds, and study molecular interactions such as the association and dissociation of complexes. It is widely applied in pharmaceutical research to characterize drugs, and in structural biology to study proteins and nucleic acids.

Photoelectron spectroscopy

Photoelectron spectroscopy, also known as ESCA (Electron Spectroscopy for Chemical Analysis), is an advanced analytical technique used primarily to study the chemical composition of surfaces and materials. Based on the interaction between photon light and surface electrons, this method provides detailed information on the elemental composition and oxidation states of atoms on the surface of samples.

Basic principles: In photoelectron spectroscopy, high-energy photons $h\nu$ are directed onto the sample surface, causing high-energy electrons (photoelectrons) to be ejected. The kinetic energy E_k of the ejected photoelectrons is measured to determine the distribution of electronic energy states in the sample. The relationship between the energy of the incident photon $h\nu$ and the kinetic energy of the photoelectrons is used to calculate the work function ϕ of the material: $h\nu = E_k + \phi$

Applications: This technique is widely used in the materials industry to characterize surface properties, assess material purity and monitor deposition and coating processes. In academic research, photoelectron spectroscopy is essential for studying catalysts, semiconductors, polymers and biological surfaces. It provides crucial information on the chemical reactivity of surfaces and contributes to the optimization of industrial processes and technological devices.

In summary, advanced spectroscopic techniques such as NMR and photoelectron spectroscopy represent essential tools for the precise analysis of molecular structures, the characterization of surfaces and the understanding of chemical interactions at the atomic scale. Their application in a wide range of fields, from structural chemistry to materials technology, testifies to their growing importance in scientific research and technological innovation.

Chapter 5: Specific applications of spectrochemistry

Spectrochemistry plays a crucial role in a wide range of applications, from industrial quality control to the study of atmospheric processes. This chapter explores how spectroscopic techniques, including group theory, interact intelligently with other methods to solve complex and varied problems.

Spectroscopy in industry

Spectroscopy is widely used in industry to ensure product quality and monitor environmental processes. Spectroscopic techniques provide detailed information on the chemical composition of materials, enabling rapid and accurate analyses that help ensure compliance with quality and safety standards.

Example: Quality control in the pharmaceutical industry

In the pharmaceutical industry, infrared (IR) spectroscopy plays an indispensable role in the quality control of raw materials, intermediates and finished products. This spectroscopic technique enables precise, non-destructive analysis of samples, providing crucial information on their molecular composition and structure.

The use of IR spectroscopy begins as soon as raw materials are received. Pharmaceutical laboratories analyze IR spectra to verify the purity of incoming compounds. This step is crucial to ensure that active ingredients comply with quality standards and are free from contaminants. For example, IR spectra can be used to identify potential impurities that could compromise the safety and efficacy of pharmaceutical products.

During the manufacturing process, IR spectroscopy is used to monitor chemical reactions and assess the progress of synthesis reactions. Laboratories can track molecular changes in intermediates in real time, ensuring consistency and quality throughout production. For example, by measuring IR spectra at different stages of synthesis, scientists can detect any unexpected formation of by-products or unwanted structural modifications.

For finished products, IR spectroscopy is used to characterize and validate their chemical composition. IR spectra serve as a unique molecular signature, enabling finished products to be compared with design specifications. This ensures that the drugs produced comply with regulatory standards and quality requirements, minimizing risks for consumers and ensuring their therapeutic efficacy.

In short, infrared spectroscopy is an essential technology in the pharmaceutical industry, helping to guarantee the quality, safety and efficacy of medicines. Its integration into quality control processes enables pharmaceutical laboratories to rapidly detect variations and contaminants, ensuring reliable pharmaceutical production in compliance with strict regulatory standards.

Atmospheric chemistry applications

Spectroscopic techniques play a crucial role in understanding atmospheric processes and monitoring pollution. They make it possible to analyze the complex interactions between chemical compounds in the atmosphere and incident light, providing essential data on pollutant concentrations and their environmental impact.

Spectroscopy, in particular nuclear magnetic resonance (NMR) spectroscopy, infrared (IR) spectroscopy and UV-Vis spectroscopy, uses group theory to interpret the spectra obtained. This theory makes it possible to analyze the molecular symmetries of atmospheric compounds and predict the spectroscopic transitions observed. For example, it helps identify chemical species present in the air, quantify their concentrations and track their chemical reactivity in the atmosphere.

In atmospheric chemistry, spectroscopy is used to study the formation and destruction mechanisms of atmospheric pollutants such as nitrogen oxides (NOx), volatile organic compounds (VOCs) and fine particles. It also helps assess the effectiveness of emission reduction strategies and monitor pollution levels in urban and rural environments.

For example, using IR spectroscopy, researchers can detect and quantify atmospheric pollutants such as carbon dioxide (CO2), carbon monoxide (CO), and polycyclic aromatic hydrocarbons (PAHs). Observed spectroscopic transitions are analyzed according to selection rules dictated by group theory, enabling precise identification of chemical species and assessment of their environmental impact.

In summary, the application of group theory to spectroscopic techniques in atmospheric chemistry enables not only the analysis of spectra and the interpretation of experimental data, but also the solution of complex problems related to atmospheric chemistry and environmental pollution. This integrated approach is essential for effective air quality management and to support global environmental sustainability efforts.

Example: Greenhouse gas monitoring

High-resolution spectroscopy, such as X-ray photoelectron spectroscopy (XPS), is used to study the interactions of greenhouse gases like carbon dioxide (CO2) with the earth's surfaces. By measuring the XPS spectra of soil and atmospheric samples, researchers can determine how these gases interact with environmental materials and influence global climate. This understanding is essential for assessing the impact of human activities on climate change and developing effective emission reduction strategies.

Integrating group theory with other spectroscopic techniques :

Group theory is an essential tool in spectrochemistry, offering a systematic approach to understanding the properties of molecules based on their symmetries, and enabling the prediction of observed spectroscopic transitions. This theory integrates significantly with other spectroscopic techniques, providing a robust mathematical framework for interpreting experimental data and understanding complex molecular interactions.

In the context of Raman spectroscopy, for example, group theory helps determine which vibrational transitions are permitted or forbidden according to Laporte's selection rules. These rules stipulate that Raman transitions are permitted if they involve a change in parity, i.e. a difference in symmetry between the initial and final states of the molecule. Representations of the molecule's symmetry group enable these transitions to be predicted by identifying the symmetries of the quantum states involved in the transition processes.

To solve these quantum mechanical problems, complex modeling software is often required. These software tools use the principles of group theory to calculate vibrational spectra and predict spectroscopic transitions with increased accuracy. They enable molecular behavior to be simulated numerically, and the results of theoretical models to be compared with experimental data obtained by Raman spectroscopy.

The application of group theory is not limited to Raman spectroscopy:

It is also crucial in infrared (IR) spectroscopy, UV-Vis spectroscopy, fluorescence spectroscopy and other techniques. In each case, it helps interpret spectra in terms of molecular symmetries, facilitating understanding of molecular structures and their interactions.

In conclusion, the integration of group theory with other spectroscopic techniques enables a more in-depth and precise analysis of molecular properties. This multidisciplinary approach is indispensable for solving complex problems in spectrochemistry, and opens up new perspectives for fundamental and applied research in many scientific and technological fields.

Integration example: Raman spectroscopy and group theory

In Raman spectroscopy, molecules react to incident light by altering their polarizability, giving rise to characteristic spectra reflecting molecular vibrational modes. Group theory plays a crucial role in the analysis and interpretation of these spectra, providing a mathematical framework for understanding molecular symmetries and predicting allowed transitions.

Raman spectra show peaks at specific frequencies corresponding to the molecule's normal modes of vibration. These modes are classified according to their symmetries under the molecule's symmetry group, determined by its structure and atomic configuration. For example, a molecule with axial symmetry may have vibration modes that respect this symmetry, such as torsion or axial deformation modes.

Group theory enables these vibrational modes to be accurately calculated and predicted by identifying the symmetry group representations that characterize each mode. The symmetries of the initial and final states involved in Raman transitions determine the permittivity of these transitions, i.e. whether they are allowed or forbidden according to selection rules. For example, a Raman transition is permitted if it preserves the overall symmetry of the molecule, while it is forbidden if it violates this symmetry.

This theoretical approach enables spectral peaks observed in Raman experiments to be precisely assigned to specific vibrational modes of the molecule under study. As a result, it enhances the validity of experimental data and facilitates their interpretation in various scientific and technological fields. In fundamental research, it helps to understand the structural and dynamic properties of molecules, while in industrial applications, it supports the development of new materials and chemicals by ensuring rigorous quality control and precise characterization of compounds.

In conclusion, this chapter highlights the versatility and power of spectroscopic techniques, supported by group theory, in fields as diverse as industry and atmospheric chemistry. These methods not only improve our understanding of chemical and environmental processes, but also facilitate the development of innovative solutions to contemporary challenges linked

to quality, the environment and sustainability.

Chapter 6: Future prospects and progress

Spectroscopy, through its many applications and advances, benefits greatly from the integration of group theory. This chapter explores how this theory continues to play a crucial role in the evolution and innovation of spectroscopic techniques, while significantly impacting various scientific and technological fields.

New developments in spectroscopy :

Recent advances in spectroscopy are intimately linked to the theoretical foundations provided by group theory. Here's how this theory enriches and supports technological advances:

- High-resolution spectroscopy: Group theory enables the precise prediction and interpretation of vibrational modes and electronic transitions observed at high resolution, facilitating the resolution of spectra and the fine characterization of samples.

- Multidimensional spectroscopy: By integrating group theory, it becomes possible to explore molecular interactions in several dimensions (time, frequency, space), offering a detailed view of dynamic processes and complex quantum states.

- Combined spectroscopy: The simultaneous use of different spectroscopic techniques is optimized by the application of selection rules based on molecular symmetries determined by group theory. This enables comprehensive analysis of samples, exploiting the complementarities of IR, Raman, NMR and other methods.

- Miniaturized spectroscopic sensors: The miniaturization of spectroscopic sensors benefits from the theoretical advances made by the groups, facilitating their integration into portable devices and constrained environments, while preserving their ability to provide precise spectroscopic information.

Implications for research and industry

Group theory not only enriches advanced spectroscopic techniques, but also has a significant impact in various fields of application:

- Medicine and biology: By enabling precise analysis of complex molecular structures, group theory supports the development of advanced biomedical diagnostics and personalized therapies based on the spectroscopic characterization of biomolecules.

- Materials and nanotechnologies: The optimization of materials and nanostructures is facilitated by group theory, which guides the design of functional materials with specific properties tailored to technological applications, such as electronics, catalysts and optical devices.

- Environmental sustainability: Spectroscopy, guided by group theory, contributes to the accurate monitoring of atmospheric pollutants and the understanding of biogeochemical processes, thus supporting environmental sustainability initiatives and the fight against climate change.

- Food and agricultural safety: Spectroscopic analysis, enhanced by group theory, is crucial for identifying food contaminants, monitoring pesticide residues and assessing the nutritional properties of foods, thus contributing to global food safety.

In conclusion, group theory continues to play an essential role in the evolution and application of advanced spectroscopic techniques. By integrating these theoretical foundations into research and industry, spectrochemistry not only extends the frontiers of scientific knowledge, but also supports the technological innovations crucial to solving the complex challenges of our time.

Conclusion

Spectrochemistry, as a crucial interdisciplinary field, relies on theoretical advances and practical applications that have shaped our understanding of molecular systems at a fundamental level. This journey through the various chapters of this book has highlighted the importance of sophisticated techniques such as infrared spectroscopy, Raman spectroscopy, UV-Vis spectroscopy, NMR, and many others, in diverse fields ranging from atmospheric chemistry to the pharmaceutical industry.

In our exploration, group theory has proved to be an essential mathematical tool for classifying quantum states and predicting spectroscopic transitions as a function of molecular symmetries. It offers a systematic approach to interpreting experimental results and understanding complex interactions within molecules. For example, it enables the determination of normal vibrational modes in IR and Raman spectra, and the prediction of permitted transitions according to selection rules.

Looking to the future, spectrochemistry continues to play a central role in developing new technologies and solving scientific challenges. Spectroscopic techniques are evolving with the emergence of new instruments and analytical methods, enabling ever finer resolution of molecular structures and more precise detection of compounds in complex matrices. Future prospects include the exploration of emerging fields such as quantum spectroscopy, high spatial resolution spectroscopy, and integration with other techniques such as microscopy and advanced computer modeling.

In conclusion, spectrochemistry remains a dynamic and fertile field for scientific research and technological innovation. Not only does it offer powerful tools for exploring the molecular world from different angles, it also contributes significantly to practical applications ranging from medicine to the environment, ensuring a lasting impact on our society and our understanding of the world around us.

List of symbols and abbreviations

Spectrochemistry: Study of the interactions between light and matter, enabling the analysis of energy transitions and molecular properties.

Quantum mechanics: Physical theory describing the behavior of subatomic particles via Schrödinger's equation.

Schrödinger equation: Fundamental equation of quantum mechanics describing the time evolution of the wave functions of quantum systems.

Symmetrical groups: Sets of transformations that preserve the symmetry properties of a molecular system.

Molecular spectra: Graphs representing the intensity of light absorbed or emitted by a molecule at different wavelengths, revealing electronic, vibrational and rotational transitions.

Vibrational spectroscopy: Technique for studying molecular vibrations by measuring energy absorption or emission in the infrared range.

Rotational spectroscopy: Analytical method analyzing energy transitions due to the rotation of molecules, typically in the microwave range.

Electron spectroscopy: Study of energy transitions associated with electrons in atomic or molecular orbitals, often in the UV-visible range.

Nuclear Magnetic Resonance (NMR): Spectroscopic technique based on the interaction of atomic nuclei with a magnetic field, used to determine molecular structure.

Group theory: Branch of mathematics studying the properties of sets of transformations (groups), applied in spectrochemistry to classify molecular symmetries and predict spectral properties.

Absorbance: Measurement of light absorption by a substance, often used in spectroscopy to quantify light-matter interaction.

Molar absorption coefficient (ε) : Constant representing a substance's ability to absorb light at a given wavelength, correlated to its concentration.

Dipole moment: A measure of the distribution of electric charges in a molecule, influencing its rotational properties and interactions with external electric fields.

Wave number (cm^{-1}): Unit in spectroscopy expressing the frequency of molecular vibrations, inversely proportional to the wavelength of light.

Beer-Lambert law: Relation describing the absorption of light by a solution as a function of the concentration of the substance, the thickness of the cell and the molar absorption coefficient.

Rotational energy (E_J): Energy associated with a molecule's rotational quantum transitions, determined by its rotational quantum number and moment of inertia.

Spherical harmonics: Solutions of Schrödinger's equation for spherically symmetrical systems, used to describe electron orbitals around the atomic nucleus.

Laguerre polynomials: Mathematical functions used to solve the radial Schrödinger equation for atoms, describing the radial distribution of the probability of the presence of an electron.

Electronic transitions: The passage of an electron from one energy level to another, often observed in electronic spectroscopy.

Transition energy: Energy required for an electron to make a transition from one quantum state to another.

Fluorescence: Emission of light by a substance after being excited by an external light source.

Fluorescence quantum efficiency: Ratio between the number of photons emitted and the number of photons absorbed during the fluorescence process.

Dipole moment: Measure of charge separation in a molecule, influencing the intensity of rotational transitions in microwave spectroscopy.

Structural chemistry: Branch of chemistry studying the structure of molecules, including molecular geometry, chemical bonds and the spatial arrangement of atoms.

Sample: Substance or material taken for analysis or experimentation in order to represent a larger population.

Electrons : Negatively charged subatomic particles present in atoms and involved in chemical bonds and reactions.

Larmor frequency: Frequency at which an atomic nucleus precessing in an external magnetic field emits or absorbs energy.

Dipole moment: Vector measuring the separation of charges in a molecule, influencing its interaction with an external electric or magnetic field.

Photoelectrons: Electrons ejected from a material under the effect of incident high-energy light (photons).

Photon: Elementary particle of light, carrying a discrete quantity of electromagnetic energy.

Nuclear precession: Rotation of an atomic nucleus around an axis under the effect of an external magnetic field.

Nuclear spin: Intrinsic property of atomic nuclei, analogous to quantized angular momentum.

NMR spectrum: Diagram showing the intensity of NMR signals as a function of the resonance frequency of atomic nuclei.

Photoelectron spectroscopy: Surface analysis technique based on the emission of photoelectrons by a material under the impact of high-energy photons.

Infrared (IR) spectroscopy: Spectroscopic technique for studying the molecular vibrations of compounds, measuring the absorption of infrared light by a molecule's chemical bonds.

IR spectrum: Graphic representation of the absorption of infrared light by a substance as a function of wavelength or wave number, with each peak corresponding to a specific vibration of molecular bonds.

Impurities: Unwanted substances in a material or product that can compromise its quality or safety.

Chemical reactions: Processes in which substances (reactants) undergo chemical transformations to form new substances (products).

Quality standards: Criteria established to ensure that products meet specified requirements in terms of purity, safety and efficacy.

Contaminants : Undesirable substances accidentally or intentionally introduced into a material or product, often in minute quantities.

References

Atkins, P., & de Paula, J. (2010). Atkins' Physical Chemistry. Oxford University Press.

Hollas, J. M. (2004). Modern Spectroscopy (4th ed.). John Wiley & Sons.

Silverstein, R. M., Webster, F. X., & Kiemle, D. J. (2004). Spectrometric Identification of Organic Compounds (7th ed.). John Wiley & Sons.

Stuart, B. (2004). Infrared Spectroscopy: Fundamentals and Applications. John Wiley & Sons.

Pavia, D. L., Lampman, G. M., Kriz, G. S., & Vyvyan, J. R. (2008). Introduction to Spectroscopy (4th ed.). Cengage Learning.

Atkins, P., & Friedman, R. (2010). Molecular Quantum Mechanics. Oxford University Press.

Herzberg, G. (1989). Molecular Spectra and Molecular Structure: I. Spectra of Diatomic Molecules. Dover Publications.

Nakamoto, K. (2009). Infrared and Raman Spectra of Inorganic and Coordination Compounds: Part A: Theory and Applications in Inorganic Chemistry. John Wiley & Sons.

Levine, I. N. (2009). Quantum Chemistry (6th ed.). Pearson Prentice Hall.

Jensen, P. (2007). Introduction to Computational Chemistry (2nd ed.). John Wiley & Sons.

Skoog, D. A., Holler, F. J., & Crouch, S. R. (2017). Principles of Instrumental Analysis (7th ed.). Cengage Learning.

Banwell, C. N., & McCash, E. M. (1994). Fundamentals of Molecular Spectroscopy (4th ed.). McGraw-Hill.

Griffiths, P. R., & de Haseth, J. A. (2007). Fourier Transform Infrared Spectrometry (2nd ed.). John Wiley & Sons.

Keeler, J. (2010). Understanding NMR Spectroscopy (2nd ed.). Wiley.

Brundle, C. R., & Baker, A. D. (1978). Introduction to Electron Spectroscopy. Academic Press.

Lindner, E., & Prigogine, I. (1961). Thermodynamics and Statistical Mechanics. Wiley.

Smith, B. C. (1999). Fundamentals of Fourier Transform Infrared Spectroscopy (2nd ed.). CRC Press.

Workman Jr, J., & Weyer, L. (2007). Practical Guide to Interpretive Near-Infrared Spectroscopy. CRC Press.

Pharmaceutical Quality Control Laboratory (2020). Handbook of Pharmaceutical Excipients. Pharmaceutical Press.

Herriott, D., & Hochstrasser, R. M. (Eds.). (1999). Infrared and Raman Spectroscopy of Biological Materials. Marcel Dekker, Inc.arcel Dekker, Inc.

Printed by Books on Demand GmbH, Norderstedt / Germany